NUMERIC

PROGRESSION

PURUSHOTTAM

KUMAR

SUMAN

Purushottam kumar suman asserts the moral right to be identified as the author of this work.

Copyright © 2019 purushottam kumar suman

All rights reserved. No part of this publication may be reproduced, distributed, or transmitted in any form or by any means, including photocopying, recording, or other electronic or mechanical methods, without the prior written permission of the publisher, except in the case of brief quotations embodied in critical reviews and certain other noncommercial uses permitted by copyright law. For permission requests, write to the publisher, addressed "Attention: Permissions Coordinator," at the mail below.

purushottamkumarsuman1002@gmail.com

ISBN: 978-17-9084-660-3

Second Edition – 10, January,2019

DEDICATED

TO MY MOM AND DAD

PREFACE

I am feeling great pleasure by presenting my first book. the book covers only a small part of sequences and series but still, it covers too many details. just like you had taken a mug of water from the ocean, which is a very small quantity but if you begin to think at an atomic level, then it becomes too magnificent.

The book is a practice to categories progression of similar type and nature. we'll get collective knowledge of the infinite number of progressions under a single topic "Numeric progression".

The book covers mainly three topics, to find nth term of any numeric progression, to find sum of nth term of any numeric progression, to find nth term from end of any numeric progression, It's the magic of mathematics that in end we'll be able to find these three by using only three formulas for any numeric progression.

Your comments are highly appreciated. Feel free to contact me for any related issues.
purushottamkumarsuman1002@gmail.com

Enjoy the magic of mathematics. All the best!

-PURUSHOTTAM KUMAR SUMAN

ACKNOWLEDGMENTS

I would like to express my gratitude toward my dad Mr. Ravindra Kumar Suman, Mom Mrs. Arti suman and sister Miss Moni kumari for helping me a lot in order to complete this book.

I would also like to thank my high school math's teacher Mr. Surendra Sinha and Science teacher Mr. Rajesh priyadarshi for boosting the curiosity in me.

Above all, I want to thank all those mathematicians who contributed to the world of sequence. A special thanks to pascal for his Pascal's triangle without which the concept of numeric progression is incomplete.

Last and not least: I beg forgiveness of all those who have been with me over the journey and whose names I have failed to mention."

PURUSHOTTAM KUMAR SUMAN

1	Sequences	1
2	Some sequences	5
3	Pascal's triangle	11
4	Numeric progression	13
	4.1 Keywords	15
	4.2 Numeric progression: what is it?	20
	4.3 N.P Kinds	23
	4.4 Sufficient no. of terms	24
	4.5 Null progression	26
	4.6 common arithmetic progression	27
	4.7 Arithmetic progression	28
	4.8 Tetranumeric progression	36
	4.9 Pentanumeric progression	48
	4.10 Hexanumeric progression	61
5	General Formulation	63
	6 Author's words	72

PURUSHOTTAM KUMAR SUMAN

1. SEQUENCES

The rising of the sun and gently setting of it, Appearance of the moon before night and gradually fading away of it till morning. This is happening in a sequence and continuing for several years. Even all planetary motions happen in a sequence. Our answer sheet in an examination is also in a unique sequence. There are a lot of examples of sequences from our day to day life but here we shall work only on sequences related directly to mathematics or from the mathematical aspects.

Any group of number are in sequence. For example: -

Seq^1 :-   1,2,3,2,4,6,4,8,12,8,16,24,...

Seq^2 :-   2,3,5,7,11,13,17,19,23,29,...

Seq^1 and Seq^2 may seems totally arbitrary in first go but after going a little deep, we can recognize a pattern for each sequence. In Seq^1 the triplets are getting double as the sequence proceed.

$$1,2,3, \quad 2,4,6, \quad 4,8,12, \quad 8,16,24,...$$
$$\quad \times 2 \quad\quad \times 2 \quad\quad \times 2$$

$$Seq^1$$

Seq^2 :- It is a set of prime numbers beginning from 2 and dots in last are indicating that this will go forever without any end. Our arbitrary lists of numbers now have a meaning.

1.1 Defining a sequence

Defining a sequence is setting procedure for the progress of a sequence term by term. Let's move further with the following terms

$$2, 5, 12, 31, 86, \ldots$$

Can we find the next term? Of course, yes! Take a pause and try to find it yourself. How many answers can we have? The answer is surprising but true! It is infinite! The given terms can be a part of an infinite number of sequences of which some are following

$$\text{Seq}^3 :\text{- } 2,5,12,31,86,249,736,2195,\ldots$$

$$\text{Seq}^4 :\text{- } 2,5,12,31,86,217,480,947,\ldots$$

$$\text{Seq}^5 :\text{- } 2,5,12,31,86,218,486,968,\ldots$$

$$\text{Seq}^6 :\text{- } 2,5,12,31,86,218,487,975,\ldots$$

Take a look at solutions below

For Seq^3,

$$\begin{array}{cccccccc} & \times 3 & \times 3 & \times 3 & \times 3 & \times 3 & \\ & +4 & +12 & +36 & +108 & +324 & +972 \\ +3 & +7 & +19 & +55 & +163 & +487 & +1459 \end{array}$$

NUMERIC PROGRESSION

$$2 \quad 5 \quad 12 \quad 31 \quad 86 \quad 249 \quad 736 \quad 2195$$

For Seq^4,

```
              +16  +16  +16  +16
          +8  +24  +40  +56  +72
       +4 +12  +36  +76 +132 +204
     +3 +7 +19 +55 +131 +263 +467
      2  5  12  31  86  217  480  947
```

The same can be done for Seq^5 and Seq^6. So, which one is perfect? Actually, all are equally reasonable. On a given sequence multiple rules can work. Also, a given sequence can belong to more than one sequence. Take another example for more clarification.

$$2, 4, 6, \ldots$$

Can we say that the sequence will proceed only as $2,4,6,8,10,12,14,16,\ldots$ not at all. It can also be $2,4,6,9,14,22,\ldots$. But what if we say sequence belonging to $2n$ where $n \in N$. Then this time we can definitely say that it will be $2,4,6,8,10,12,14,16,\ldots$. As several rules can be working on a single sequence but a single rule can arise only one sequence. That's why we will prefer defining the rule of the sequence first and then proceed it further according to that. However, our approach can be two-directional, we can go

(i) Sequence ⟹ Observation ⟹ Defining rule

or

(ii) Rule ⟹ Sequence.

We will use it throughout the book. The study of sequences is completely a study of choice. We will choose the rule of our own to define sequences as we want to see +

-the sequences.

We can represent the Defining rule in multiple ways but again we will choose the representation of our choice. Which is

- In a sentence, mostly for those sequences which is hard to represent in equation forms or other forms like a sequence of all prime numbers from 1.
- In equations, like $S(n) = n^2 - n - 1$. Putting integers(n) in the equation will give the terms. Let's try this out, start putting natural numbers i.e. 1,2,3,4,5,…

    $S(1) = -1$ $S(4) = 11$

    $S(2) = 1$ $S(5) = 19$

    $S(3) = 5$ $S(6) = 29$

    The sequence will be -1,1,5,11,19,29,…

2. SOME SEQUENCES

❖ 2.1 Arithmetic sequences

The progression in which the difference of two consecutive terms is a constant (also known as common difference) are known as arithmetic progression.

EX-   2,4,6,8,10,…

1,2,3,4,5,6,…

Set of Natural number or even number

If the initial term of an arithmetic progression is A_1 and the common difference of consecutive terms is d, then the *n*th term of the sequence is given by:

$$T_n = A_1 + (n-1)d$$

❖ 2.2 Lucas numbers

2, 1, 3, 4, 7, 11, 18, 29, 47, 76, …

Defined by

$$L(n) = L(n-1) + L(n-2) \text{ for } n \geq 2,$$

with $L(0) = 2$ and $L(1) = 1$.

❖ 2.3 Fibonacci numbers

0, 1, 1, 2, 3, 5, 8, 13, 21, 34, ...

$F(n) = F(n-1) + F(n-2)$ for $n \geq 2$,

with $F(0) = 0$ and $F(1) = 1$.

❖ 2.4 Tribonacci numbers

0, 1, 1, 2, 4, 7, 13, 24, 44, 81, ...

$T(n) = T(n-1) + T(n-2) + T(n-3)$ for $n \geq 3$,

with $T(0) = 0$ and $T(1) = T(2) = 1$.

❖ 2.5 Pell numbers

0, 1, 2, 5, 12, 29, 70, 169, 408, 985, ...

$a(n) = 2a(n-1) + a(n-2)$ for $n \geq 2$,

with $a(0) = 0$, $a(1) = 1$.

❖ 2.6 Jacobsthal numbers

0, 1, 1, 3, 5, 11, 21, 43, 85, 171, 341, ...

$a(n) = a(n-1) + 2a(n-2)$ for $n \geq 2$,

with $a(0) = 0$, $a(1) = 1$.

NUMERIC PROGRESSION

❖ 2.7 Perrin numbers

3, 0, 2, 3, 2, 5, 5, 7, 10, 12, …

$P(n) = P(n-2) + P(n-3)$ for $n \geq 3$,

with $P(0) = 3$, $P(1) = 0$, $P(2) = 2$.

❖ 2.8 Prime numbers

2, 3, 5, 7, 11, 13, 17, 19, 23, 29, …

with $n \geq 1$.

❖ 2.9 Factorials n!

1, 1, 2, 6, 24, 120, 720, 5040, 40320, 362880, …

$n! := 1 \cdot 2 \cdot 3 \cdot 4 \cdot \cdots \cdot n$ for $n \geq 1$,

with $0! = 1$ (empty product)

❖ 2.10 Sylvester's sequence

2, 3, 7, 43, 1807, 3263443, 10650056950807,
113423713055421844361000443, …

$a(n + 1) = a(n) \cdot a(n - 1) \cdot \cdots \cdot a(0) + 1 = a(n)^2 - a(n) + 1$ for $n \geq 1$,

with $a(0) = 2$.

❖ 2.11 Fermat numbers

3, 5, 17, 257, 65537, 4294967297, 18446744073709551617, 340282366920938463463374607431768211457, …

$$F_n = 2^{2^n} + 1 \text{ for } n \geq 0.$$

❖ 2.12 Cullen numbers

1, 3, 9, 25, 65, 161, 385, 897, 2049, 4609, 10241, 22529, 49153, 106497, …

$$C_n = n \cdot 2^n + 1, \text{ with } n \geq 0.$$

❖ 2.13 Pronic numbers

0, 2, 6, 12, 20, 30, 42, 56, 72, 90, …

$$2t(n) = n(n+1), \text{ with } n \geq 0.$$

❖ 2.14 Woodall numbers

1, 7, 23, 63, 159, 383, 895, 2047, 4607, …

$$n \cdot 2^n - 1,$$

with $n \geq 1$.

❖ 2.15 Stella octangular numbers:

0, 1, 14, 51, 124, 245, 426, 679, 1016, 1449, 1990, 2651, 3444, 4381, ...

$n(2n^2 - 1)$, with $n \geq 0$.

❖ 2.16 Triangular numbers

0, 1, 3, 6, 10, 15, 21, 28, 36, 45, ... $t(n) = C(n+1, 2) = n(n+1)2 = 1 + 2 + \cdots + n$ for $n \geq 1$,

with $t(0) = 0$ (empty sum).

❖ 2.17 Square numbers (n^2)

0, 1, 4, 9, 16, 25, 36, 49, 64, 81, ...

$n^2 = n \times n$

❖ 2.18 Cube numbers n^3

0, 1, 8, 27, 64, 125, 216, 343, 512, 729, ...

$n^3 = n \times n \times n$

❖ 2.19 Decimal expansion of π

3, 1, 4, 1, 5, 9, 2, 6, 5, 3, ...

The ratio of a circle's circumference to its diameter

❖ 2.20 Decimal expansion of e

$$2, 7, 1, 8, 2, 8, 1, 8, 2, 8, \ldots$$

Euler's number in base 10.

❖ 2.21 Decimal expansion of the golden ratio φ

$$1, 6, 1, 8, 0, 3, 3, 9, 8, 8, \ldots$$

$= (1 + \sqrt{5})/ 2 = 1.6180339887\ldots$ in base 10.

❖ 2.22 Narayana's cows

$$1, 1, 1, 2, 3, 4, 6, 9, 13, 19, \ldots$$

The number of cows each year if each cow has one cow a year beginning its fourth year.

❖ 2.23 Lazy caterer's sequence

$1, 2, 4, 7, 11, 16, 22, 29, 37, 46, \ldots$

The maximal number of pieces formed when slicing a pancake with n cuts.

3 PASCAL'S TRIANGLE

```
                    1
                  1   1
                1   2   1
              1   3   3   1
            1   4   6   4   1
          1   5  10  10   5   1
        1   6  15  20  15   6   1
      1   7  21  35  35  21   7   1
    1   8  28  56  70  56  28   8   1
  1   9  36  84 126 126  84  36   9   1
1  10  45 120 200 252 200 120  45  10   1
```

Can you observe what's going on here? Each number is the numbers directly above it added together. For example, in row 4, 3=1+2. That's how this will grow. We will consider the sequence diagonally as mentioned in the figure below.

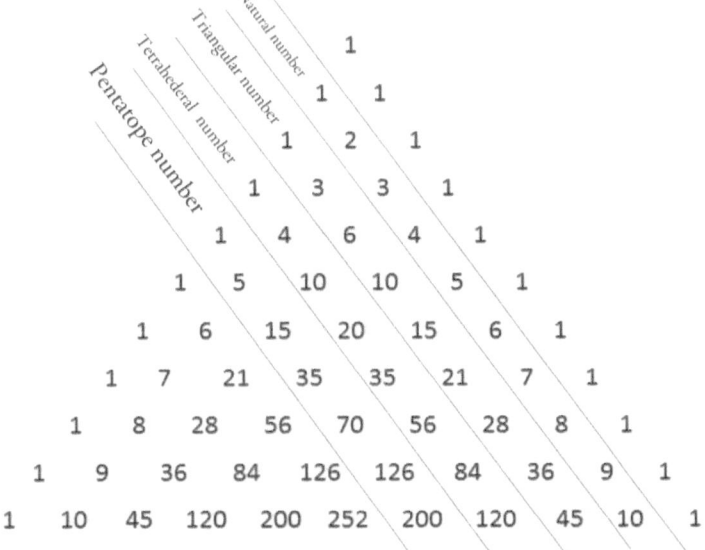

- ❖ The nth term of natural number = n
- ❖ Nth terms of a triangular number
$$\frac{n(n+1)}{2}$$
- ❖ Nth terms of tetrahedral number
$$\frac{n(n+1)(n+2)}{3!}$$
- ❖ Nth terms of pentatope number
$$\frac{n(n+1)(n+2)(n+3)}{4!}$$

And so on ……

4. NUMERIC PROGRESSION

In previous sections, we have gone through several types of sequences, series and pascal's triangle. In this section, we will utilize all that knowledge, specially Pascal's triangle. Remind the triangular number from pascals' triangles. which is

$$0, 1, 3, 6, 10, 15, 21, 28, 36, 45, \ldots$$

Now start writing the difference of two consecutive terms slightly above and in between the respective pairs as shown below.

$$\begin{array}{ccccccccc} & +1 & +2 & +3 & +4 & +5 & +6 & +7 & \\ 0 & & 1 & & 3 & & 6 & & 10 & & 15 & & 21 & & 28 \end{array}$$

We found that the set of differences are natural numbers, which is an arithmetic progression. Do the same steps for tetrahedral number i.e. $0, 1, 4, 10, 20, 35, 56, 84, 120, \ldots$

$$\begin{array}{ccccccccc} & +1 & +3 & +6 & +10 & +15 & +21 & +28 & \\ 0 & & 1 & & 4 & & 10 & & 20 & & 35 & & 56 & & 84 \end{array}$$

This time go, one more step above.

$$\begin{array}{ccccccccc} & & +2 & +3 & +4 & +5 & +6 & +7 & \\ & +1 & +3 & +6 & +10 & +15 & +21 & +28 & \\ 0 & & 1 & & 4 & & 10 & & 20 & & 35 & & 56 & & 84 \end{array}$$

what we can observe is set of differences of tetrahedral number are in progression whose set of differences are in arithmetic progression. Again do the same steps with pentagonal numbers i.e. 1,5,15,35,70,…

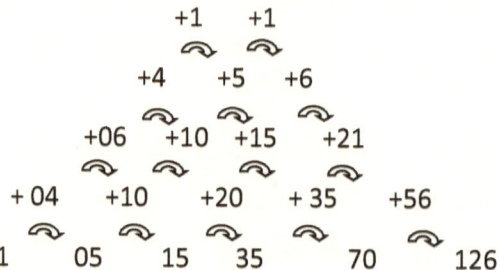

What we found is the set of differences are in progression, whose sets of differences are in progression, whose sets of differences are arithmetic progression. Isn't is too confusing? Don't worry! We'll have a completely simpler concept of it. Before this take a look at some important keywords.

NUMERIC PROGRESSION

4.1. KEYWORDS

➢ Principle progression

Any progression on which we are working is principle progression. It is just a term used to recognize the given progression. We are free to choose any progression as principle progression. If we are working with the triangular number then the principle progression is 1,3,6,10,15,21,... If we are working on Fibonacci numbers then the principle progression is 0,1,1,2,3,5,8,13,21,...

Ex:-

```
                    +2     +2     +2
            +2    +4     +6     +8
       +1   +0 3  +07    +13    +21
   2    03    06    13     26     47

       Seq 5.1.1      principle progression
```

```
                    +0   +0
              +1    +1    +1
        +2    +3    +4    +5
    8    10    13    17    22

       Seq 5.1.2      principle progression
```

15

> Explanation of progression

Expressing principle progression in additive or subtractive forms in layers until it approaches topmost row with only one term is known as explaining the progression. For explanation write the given principle progression with some gaps between the terms. Beginning from the first term draw curve arrow pointing toward the next term. In the gaps slightly above of all arrows write the difference of two consecutive terms with appropriate signs i.e. +/-. Now do the process again for the set of terms that appeared now. Do this process until you reach topmost row. In each step there will be a loss on one term.i.e. if you had taken five terms, then one step up you will be left with four terms, then three, then two and finally one. Actually, by explaining the progression, we will come to know about the nature of the progression. In the case of numeric progression, the topmost will always be zero (stay tuned to know more about it). While in the rest of cases we'll be left with a constant.

```
                              +0
                          +1      +1
                       +2     +3      +4
Principle            +8   +10   +13   +17
progression   →    2    10    20    33    50
```

Sequence 5.1.3

NUMERIC PROGRESSION

```
                    -16
                   ⌒
              +8       -8
             ⌒       ⌒
         -4      +4      -4
        ⌒      ⌒      ⌒
     +2     -2     +2     -2              Principle
    ⌒    ⌒     ⌒    ⌒
  -1     1    -1     1    -1  ──────▶     progression
```

Sequence 5.1.4 : Explanation of first five terms of $(-1)^n$

> Slope

In an explained progression each row of differences will be called as the slope. It is also a term just like principle progression used to represent each row. Unlike principle progression, we are not free to name any slope as our choice. Counting of the slope will always begin from topmost row so as to prevent its variation with the principle progression. Name the topmost slope with zero as the only element as slope 1 and increase the integral value as you go down. If a progression has a single finite value at the top then it is beginning from slope 2.

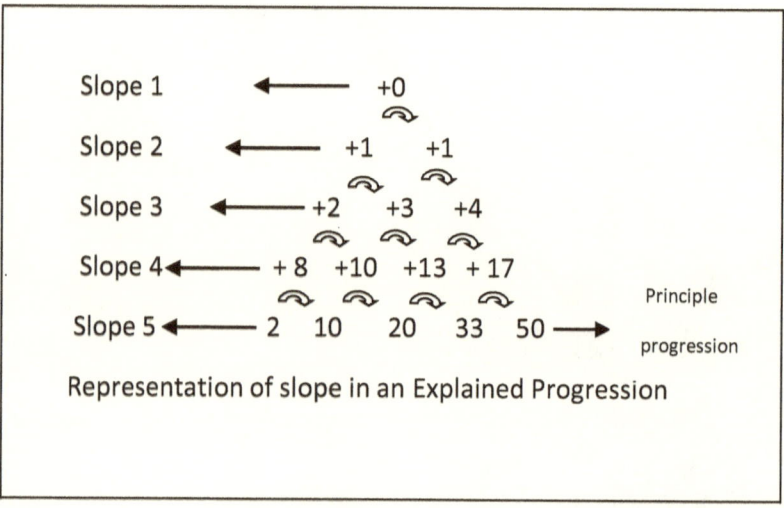

Representation of slope in an Explained Progression

NUMERIC PROGRESSION

> A_n

The first term of any slope will be denoted by A with subscript number of the slope.

- A_1 is the first term of slope 1. which is always 0 or a constant.

- A_2 is the first term of slope 2.

- A_3 is the first term of slope 3.

- A_4 is the first term of slope 4.

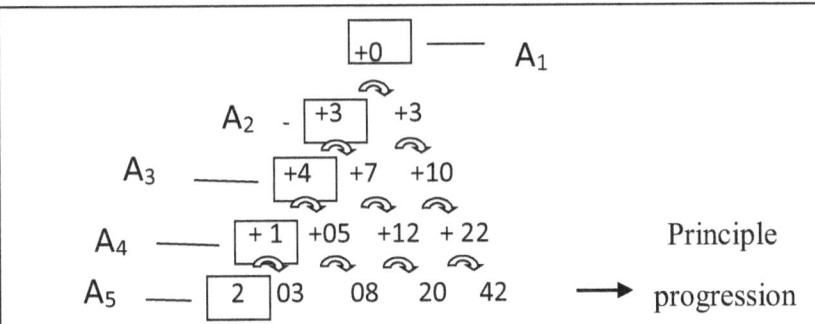

Representation of A_n in an explained Progression

4.2. NUMERIC PROGRESSION: WHAT IS IT?

Any sequence of number which on explanation approaches slope 1 of which the only element is zero are in numeric progressions.

Let's start with an integer 2. Remember the whole process is completely choice based. You are independent to choose any integer as the first terms of any slope. Again, you are independent to choose + or − signs and number of terms (we are taking 5 terms here). Write it as below.

$$+2 \quad +2 \quad +2 \quad +2 \quad +2$$

Again, take an integer 1. Place 1 slightly left and below 2. Begin adding the terms as 1+2=3 then 3+2=5 and so on.

Remember this will be slope 3. As the difference between each 2 is 0 which is an element of slope 1 and 2,2,2,2,2,… will belongs to slope 2. In this slope, we'll get six terms which are one more than 5.

$$\begin{array}{ccccccc} & +2 & +2 & +2 & +2 & +2 \\ +1 & +3 & +5 & +7 & +9 & +11 \end{array}$$

Do the process again with 3 as first term. We'll get

NUMERIC PROGRESSION

```
        +2    +2    +2    +2    +2
     +1    +3    +5    +7    +9    +11
   3    4    7    12    19    28    39
```

We can continue the process as long as we want. This will continue to give different types of progression of different slopes. Can you feel any similarity with Pascal's triangles? Of Couse yes! The difference is that in Pascal's triangle first term was fixed i.e.1 But here are we are choosing different integers as the first terms. In this process, we went to the top to down by assuming different integers as first terms. Let's move from bottom to top. Have a look at squares of natural numbers.

$$1, 4, 9, 16, 25, 36, 49, 64, 81, \ldots$$

Explain it

```
            0     0     0     0
         +2   +2    +2    +2    +2
      +3   +5    +7    +9    +11    +13
    1    4    9    16    25    36    49
```

Clearly, it is also in numeric progression. Since the topmost slope contains only zero. Do the same process for the first four terms of "Lazy caterer's sequence", first five terms of "cubes of natural numbers and Stella octangular number".

Take reference for sequences from section 2. Not surprisingly they are also in numeric progression.

Numeric progression contains the infinite number of progressions only of different slopes. Even in each slope, there is an infinite number of progressions. It's all starts with 0,0,0,0,… then the next slope contains common elements. After it next slopes belong to Arithmetic progression and rest will be named in a moment.

In conclusion, if you had some terms of a sequence then explain it if it approaches slope 1, then you can do all of the upcoming things with it. In upcoming sections, we'll see its kind, how to find n terms of a particular numeric progression? How to find the sum of n terms of a particular numeric progression? How to find nth term from last of numeric progression? We shall perform a lot of problems to have a clear-cut vision about numeric progressions. So, gear up your mind to go on a long drive in the numeric progression world.

4.3. N.P. KINDS

Let's go from top to bottom. Slope 1 has only one element which is zero so, we shall call it as the Null progression. Slope 2 is created from a constant, it will be known as the common arithmetic progression since the common differences are 0. The Element of slope 2 is also called as d (where d≠0) in common usage. Next to it naming will proceed further by adding a number of slopes as a suffix to numeric progression. For the example: - progression belonging to slope 3 will be named as trinumeric progression, slope 4 as tetranumeric progressions, slope 5 as pentanumeric progression and so on. Below is the chart.

Null progression

Common arithmetic progression or Dinumeric progression

Arithmetic progression or Trinumeric progression

Tetranumeric progression

Pentanumeric progression

Hexanumeric progression

Septanumeric progression

Up to infinity…

4.4. SUFFICIENT NUMBER OF TERMS

The minimum number of terms required to identify which type of progression is the given principle progression is the sufficient number of terms.

Any progression should always reach to slope 1 on the explanation for its confirmation as the numeric progression. If it is not approaching then it is insufficient progression. In section 5.2. we get to know that the number of terms decreases continuously by 1 as we go from bottom to top. With any progression, there are only two chances either it will be left with a constant or zero at the top. If there is a single constant at the top then the slope 2 will definitely have 2 terms and we need at least three terms to confirm a progression as arithmetic progression. So below it all will be completely vague to determine.

Sequence 5.4.1	Sequence 5.4.2 +0
+2	+2 +2
+3 +5	+3 +5 +7
2 5 10	2 5 10 +17
An insufficient progression	A sufficient progression

In sequence 5.4.2 there are four terms in principle progression. That's why it approaches slope 1 while in sequence 5.4.1, we left with a constant resulting in an insufficient progression. Ask the question to yourself. Of which sequence we can predict the upcoming terms? The answer is clearly 5.4.2. See the list below of the sufficient number of terms for each kind of numeric progression.

- Three for Arithmetic progression
- Four for Tetranumeric progression
- Five for Tetranumeric progression
- Six for Hexanumeric progression

And so on…

4.5. NULL PROGRESSION

A progression whose all terms are zero is called a null progression. The set of differences of common arithmetic progression will always have one element i.e. 0.

Null progression: - 0,0,0,0,0,0,…

Slope 1 of every Numeric progression is null progression.

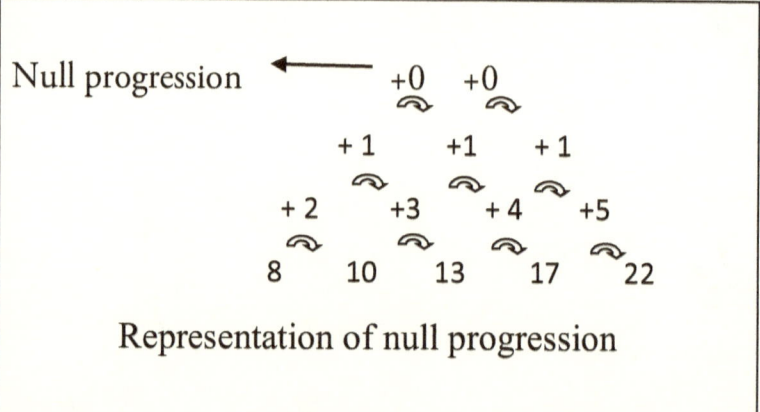

Representation of null progression

4.6. COMMON ARITHMETIC PROGRESSION

A progression whose all terms are constant is called as Common arithmetic progression. Slope 2 of Numeric progression is Common arithmetic progression. The element of slope 2 is popularly known as a constant or common difference. Denoted by d. It can be any integer other than zero, since 0,0,0,0,0, 0,… belongs to slope1.

Ex: - 1,1,1,1,1,…

2,2,2,2,2….

Nth term

Is always A_1

Summation

Sum of nth term = nA_1

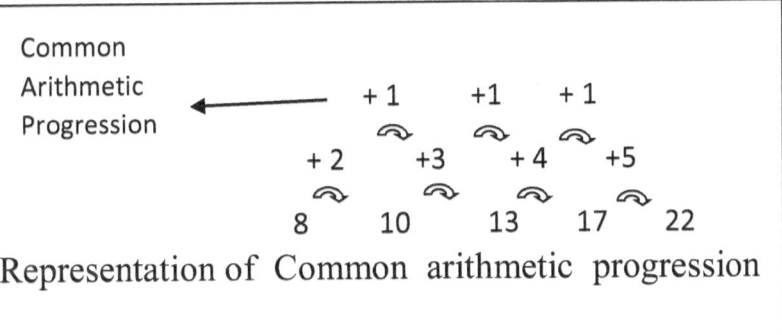

Representation of Common arithmetic progression

4.7. ARITHMETIC PROGRESSION

The progression in which the difference of two consecutive terms is a constant (also known as common difference) are known as arithmetic progression, can also be called as Trinumeric progression.

EX-   2,4,6,8,10,...

   1,2,3,4,5,6,... or

Set of natural numbers, set of even numbers beginning from 2, Set of odd numbers beginning from 1, etc.

Let's see what's happening in the arithmetic progression. Take two integers. Use one as the first term and start adding the second integer as a common difference. For instance, here we are taking 3 and 2.

$T_1 = A_1 = 3$

$T_2 = A_1 + d = 3+2 = 5$

$T_3 = T_2 + d = 5+2 = 7$

$T_4 = T_3 + d = 7+2 = 9$

Arithmetic Progression

3   +2   5   +2   7   +2   9   +2   11

Representation of Arithmetic progression

That's how arithmetic progression proceeds. Keeping adding the constant to the terms and you will keep going further. Let's work out on how to find its nth term.

4.7.1 Nth TERM OF ARITHMETIC PROGRESSION

Finding nth term isn't so tricky, it is easy and simply predictable. Just use the variables in place of the numbers and follow the same procedures. Below is the skeleton structure of an arithmetic progression.

$$\begin{array}{cccccc}
 & +0 & +0 & +0 & +0 & \\
 & +A_2 & +A_2 & +A_2 & +A_2 & +A_2 \\
A_3 & T_2 & T_3 & T_4 & T_5 & T_6
\end{array}$$

Skelton Structure

Terms used

- A_3 for the first term of Arithmetic progression
- T_n for terms of slope 3.
- A_2 for the common difference.

Thus, we can express given Arithmetic Progression in the following manner

$T_1 = A_3$

$T_2 = T_1 + A_2 = A_3 + A_2$

$T_3 = T_2 + A_2 = A_3 + 2A_2$

$T_4 = T_3 + A_2 = A_3 + 3A_2$

$T_5 = T_4 + A_2 = A_3 + 3A_2$

$T_6 = T_5 + A_2 = A_3 + 4A_2$

Do the simple observation, From the given set we can conclude that A_3 is common in all expressions of arithmetic progressions. The coefficient of A_2 is one less than the number of terms i.e. for T_3 it is 2, T_4 it is 3 and so on. Or more simply in the pattern (n-1). Now assemble all of these to find the nth term of an arithmetic progression.

$$T_n = A_3 + (n-1)A_2$$

Use this formula to answer the following questions: -

To identify the Ap

4.7.2. ARITHMETIC SERIES: To find the sum of the nth term of an arithmetic progression

Consider the case, if someone gave you 3 balls and after it, he gives you two balls each day for three days. With how many balls you will be left at the end? This requires only a little calculation. But what if I ask you after the end of 100 days, ...after the end of a thousand days. Although it seems too complex but not so. In this section, we shall learn an easy method for this.

$$
\begin{array}{ccccccc}
 & & +0 & +0 & +0 & +0 & \\
 & +A_2 & +A_2 & +A_2 & +A_2 & +A_2 & \\
A_3 & T_2 & T_3 & T_4 & T_5 & T_6 \\
\end{array}
$$

Skelton Structure

Terms used

- A_3 for first the term of Arithmetic progression
- T_n for terms of slope 3.
- A_2 for the common difference.

At the end of day 1, you will have three balls means A_3. And next day we'll have 3+2=5 balls, the day after 5+2=7 balls and so on… lets again replace the digit with variables and you will find the expression sum as

For,

$s_1 = A_3$

$s_2 = A_3 + T_2 = 2A_3 + A_2$

$s_3 = A_3 + T_2 + T_3 = 3A_3 + 3A_2$

$s_4 = A_3 + T_2 + T_3 + T_4 = 4A_3 + 6A_2$

$s_5 = A_3 + T_2 + T_3 + T_4 + T_5 = 5A_3 + 10A_2$

$s_6 = A_3 + T_2 + T_3 + T_4 + T_5 + T_6 = 6A_3 + 15A_2$

now,

We can write it as

$nA_3 + vA_2$ …….(1)

where n is the number of terms, we are adding and v can be expressed as

For,

$s_1 = 0$ $s_2 = 1$ $s_3 = 3$

$s_4 = 6$ $s_5 = 10$ $s_6 = 15$

which is a triangular number of which we are already familiar. It is beginning from 0. So, its nth term will be

$$\frac{n(n-1)}{2}$$

Now replacing the value of v in equation 1st. we will get

$$S_n = nA_3 + \frac{n(n-1)}{2} \times A_2$$

or it can be more simplified as

$S_n = n\backslash 2\ [2A_3 + (n-1)A_2]$

Now try to work on the problem of the ball with the help of this formula. You can calculate it in a while.

4.7.3. TO FIND N^{TH} TERM FROM THE END OF AN ARITHMETIC PROGRESSION

Let's go reverse of the flow. We know that arithmetic sequences proceed by adding a constant d. But what if we reverse the sequence? will it be still in arithmetic progression? Yes! Of course, but this time we will have to keep subtracting the constant rather than adding it. More simply it will be an arithmetic progression with common difference -d and with given the last term as the first term. Consider the Skelton structure of Arithmetic progression

$$
\begin{array}{cccccc}
& +0 & +0 & +0 & +0 & \\
& +A_2 & +A_2 & +A_2 & +A_2 & +A_2 \\
A_3 & T_2 & T_3 & T_4 & T_5 & T_6
\end{array}
$$

and rewrite it as

$$
\begin{array}{cccccc}
& +0 & +0 & +0 & +0 & \\
& -A_2 & -A_2 & -A_2 & -A_2 & -A_2 \\
L \text{ or } T_6 & T_5 & T_4 & T_3 & T_2 & A_3
\end{array}
$$

The general formula of an arithmetic progression is

$$T_n = A_3 + (n-1)D$$

Replace A_3 with L and D with $-D$

$T_n = L-(n-1)D$

Where

- L is the last term
- n is used for the number of terms from the end which we have to find
- D is for the common difference.

4.8. TETRANUMERIC PROGRESSION

Progressions whose sets of differences are in the form of an arithmetic progression are Tetranumeric progression. Explain the squares of natural numbers. We will found a tetranumeric progression with common difference 2, $A_3=3$ $A_4=1$.

$$
\begin{array}{ccccccccc}
 & & +2 & & +2 & & +2 & & +2 \\
 & +3 & & +5 & & +7 & & +9 & & +11 \\
1 & & 4 & & 9 & & 16 & & 25 & & 36
\end{array}
$$

Seq.5.7.1: Sequence of squares of natural number, $S(n) = n^2$

Examples are infinite including triangular number given by pascal. Make your own tetranumeric progression and verify the upcoming theory for that. Some have been done for you as a reference.

Ex – 2,4,7,11,16,22,…

4,7,12,19,28,39,52,…

2,5,10,17,26,37,…

NUMERIC PROGRESSION

Some explained Tetranumeric progression

```
        +1   +1   +1
       ⌒    ⌒    ⌒
      -7   -6   -5   -4
     ⌒    ⌒    ⌒    ⌒
   28   21   15   10   06
```
Sequence 5.7.2

```
       +2    +2    +2    +2
      ⌒     ⌒     ⌒     ⌒
    +3   +5    +7    +9    +11
   ⌒    ⌒    ⌒    ⌒    ⌒
  2    5    10   17   26   37
```
Sequence 5.7.3

```
       +3    +3    +3    +3
      ⌒     ⌒     ⌒     ⌒
    +3   +6    +9   +12   +15
   ⌒    ⌒    ⌒    ⌒    ⌒
  2    5    11   20   32   47
```
Sequence 5.7.4

```
       +1    +1    +1    +1
      ⌒     ⌒     ⌒     ⌒
    +3   +4    +5    +6    +07
   ⌒    ⌒    ⌒    ⌒    ⌒
  5    8    12   17   23   30
```
Sequence 5.7.5

4.8.1 Nth TERM OF TETRANUMERIC PROGRESSION

Take any arbitrary integer P and an arithmetic progression say $Z_1, Z_2, Z_3, Z_4, Z_5, Z_6$... Start the addition of integer and terms of the arithmetic progression. This will birth a tetranumeric sequence.

$$\begin{array}{ccccccc}
& +0 & +0 & +0 & +0 & & \\
& +A_2 & +A_2 & +A_2 & +A_2 & +A_2 & \\
A_3 & Z_2 & Z_3 & Z_4 & Z_5 & Z_6 & \\
A_4 & T_2 & T_3 & T_4 & T_5 & T_6 & T_7
\end{array}$$

Skelton structure of Tetranumeric progression

Where,

A_4 is the first term of Tetranumeric progression

A_3 is the first term of slope 3.

A_2 or D is the common difference.

and

$Z_2 = A_3 + A_2$

$Z_3 = A_3 + 2A_2$

$Z_4 = A_3 + 3A_2$

$Z_5 = A_3 + 3A_2$

$Z_6 = A_3 + 4A_2$

NUMERIC PROGRESSION

Thus, we can express given tetranumeric Progression in the following manner

$T_1 = A_4$

$T_2 = A_4 + A_3$

$T_3 = T_2 + Z_2 = A_4 + 2A_3 + d$

$T_4 = T_3 + Z_3 = A_4 + 3A_3 + 3d$

$T_5 = T_4 + Z_4 = A_4 + 4A_3 + 6d$

$T_6 = T_5 + Z_5 = A_4 + 5A_3 + 10d$

From the given set we can conclude that A_4 is common in all terms of tetranumeric progression, The coefficient of A_3 is in the pattern of (n-1) And Coefficient of D appears in sequence 0,0,1,3,6,10,15,…

We already know about the sequence, its triangular number beginning two places later. We will just subtract one in the formula of the nth term of a triangular number. Which is

$$\frac{(n-2)(n-1)}{2}$$

Now we can assemble all these to find the nth term of tetranumeric progression.

$$T_n = A_4 + (n-1)A_3 + \frac{(n-2)(n-1)D}{2}$$

or

$$T_n = A_4 + (n-1)[A_3 + \frac{(n-2)D}{2}]$$

4.8.2 TETRANUMERIC SERIES: To find the sum of the nth term of tetranumeric progression.

So far, we know a lot of progressions including $S(n)=n^2$ and Triangular numbers, which is in tetranumeric progression. Now the challenge is to find its sum up to the nth term. Consider the Skelton structure of Tetranumeric progression

$$
\begin{array}{ccccccc}
 & +0 & +0 & +0 & +0 & & \\
 & +A_2 & +A_2 & +A_2 & +A_2 & +A_2 & \\
A_3 & Z_2 & Z_3 & Z_4 & Z_5 & Z_6 & \\
A_4 & T_2 & T_3 & T_4 & T_5 & T_6 & T_7
\end{array}
$$

Sum Can be expressed as

For,

$s_1 = A_4$

$s_2 = A_4 + T_2 = 2A_4 + A_3$

$s_3 = A_4 + T_2 + T_3 = 3A_4 + 3A_3 + d$

Samewise

$s_4 = 4A_4 + 6A_3 + 4d$

$s_5 = 5A_4 + 10A_3 + 10d$

$s_6 = 6A_4 + 15A_3 + 35d$

now,

it can be expressed as

$nA_4 + vA_2 + \#D$.......(1)

where n is the number of terms, v can be expressed as

For,

$s_1 = 0$ $s_2 = 1$ $s_3 = 3$

$s_4 = 6$ $s_5 = 10$ $s_6 = 15$

making progression 0,1,3,6,10,... whose nth term is

$$\frac{n(n-1)}{2}$$

And # can expressed as

For,

$s_1 = 0$ $s_2 = 0$ $s_3 = 1$

$s_4 = 4$ $s_5 = 10$ $s_6 = 20$

making progression 0,0,1,4,10,20,...(Tetrahedral number beginning one place later) so we'll subtract one from n.

Before

$$\# = S(n-1) = \frac{(n-2)(n-1)n}{6}$$

Now replacing the value of v and # in eq^n 1st

NUMERIC PROGRESSION

$$S_n = nA_4 + \frac{n(n-1)A_3}{2} + \frac{(n-2)(n-1)nD}{6}$$

Or $S_n = \frac{1}{6}[nA_4 + 3n(n-1)A_3 + (n-2)(n-1)n]$

4.8.2.1 SUM OF SQUARES OF NATURAL NUMBERS

$$\begin{array}{cccc} +2 & +2 & +2 & +2 \\ +3 \quad +5 \quad +7 \quad +9 \quad +11 \\ 1 \quad 4 \quad 9 \quad 16 \quad 25 \quad 36 \end{array}$$

Sequence of squares of natural number, $S(n) = n^2$

$A_2 = 2$, $A_3 = 3$, $A_4 = 1$

Put this in the formula of sum of n terms of tetranumeric progression. We shall get

$$S_n = nA_4 + \frac{n(n-1)A_3}{2} + \frac{(n-2)(n-1)nD}{6}$$

$\sum n^2 = n + 3n(n-1)/2 + 2(n-2)(n-1)/6$

4.8.3 n^{th} TERM FROM END

Consider the Skelton structure of Tetranumeric progression

$$\begin{array}{cccc} +0 & +0 & +0 & +0 \\ +A_2 \curvearrowright +A_2 \curvearrowright +A_2 \curvearrowright +A_2 \curvearrowright +A_2 \\ A_3 \curvearrowright Z_2 \curvearrowright Z_3 \curvearrowright Z_4 \curvearrowright Z_5 \curvearrowright Z_6 \\ A_4 \curvearrowright T_2 \curvearrowright T_3 \curvearrowright T_4 \curvearrowright T_5 \curvearrowright T_6 \curvearrowright T_7 \end{array}$$

Now, rearranged it as

$$\begin{array}{cccc} +0 & +0 & +0 & +0 \\ +A_2 \curvearrowright +A_2 \curvearrowright +A_2 \curvearrowright +A_2 \curvearrowright +A_2 \\ -Z_6 \curvearrowright -Z_5 \curvearrowright -Z_4 \curvearrowright -Z_3 \curvearrowright -Z_2 \curvearrowright -A_3 \\ L/T_7 \curvearrowright T_6 \curvearrowright T_5 \curvearrowright T_4 \curvearrowright T_3 \curvearrowright T_3 \curvearrowright A_4 \end{array}$$

Inverse skelton structure

The progression is still in tetranumeric progression. The general formula for the nth term of Tetranumeric progression is

$$T_n = A_4 + (n-1)\left[A_3 + \frac{(n-2)D}{2}\right]$$

replace A_4 with L (or last term) and A_3 with $-\emptyset$ as it is in the new inverse skeleton structure of tetranumeric progeression.

where ø stands for the first term of slope 3 (in this case it is Z_6). terms of slope 3 turned negative as we are moving from the backside.

now,

Replacing the values

$$T_n = L + (n-1)[-ø + \frac{(n-2)D}{2}]$$

Finding ø

ø is the last term of slope 3 (arithmetic progression). simply we can use the formula of the nth term of arithmetic progression to find this. we can see no of terms in slope 3 is one less than the number of terms in slope 4. so we will subtract one from the N of the general formula of arithmetic progression to make formula fit for work with teranumeric progression

$$T_n = A_3 + (N-1)d$$

$$ø = T_n(new) = A_3 + (N-2)d$$

where N is the number of terms in tetranumeric progression.

Now putting $A_3+(N-2)d$ in place of ø

$$T_n = L + (n-1)[-\{A_3+(N-2)d\} + \frac{(n-2)D}{2}]$$

NUMERIC PROGRESSION

$$T_n = L - (n-1)\left[A_3 + (N-2)d - \frac{(n-2)D}{2}\right]$$

Where

L is last term

n is used for the number of terms from the end which we have to find

A_3 is used for the first term of the Third slope

D is for the common difference of the second slope

And N is for the number of terms in sequence L.

4.9 PENTANUMERIC PROGRESSION

The progression whose differences are in the tetranumeric progression is pentanumereic progression. It belongs to slope 5 of Numeric progression. Cubes of the natural numbers are in pentanumeric progression. Some examples are

$$2,6,11,19,32,\ldots$$

$$3,5,9,16,27,\ldots$$

$$1,5,12,24,43,\ldots$$

Some Explained Pentanumeric Progression

```
                    +0
                 +3     +3
              +4    +7    +10
           +1   +05   +12   +22      Principle
           2   03    08   20   42  → progression
              Sequence 5.8..1
```

48

NUMERIC PROGRESSION

```
              +0
            ⌒
        +2     +2
        ⌒    ⌒
      +3   +5   +07
      ⌒   ⌒   ⌒
    +1  +04  +09  +16
    ⌒   ⌒   ⌒   ⌒
   2  03   07   16   32
```

Sequence 5.8..2

```
              +0
            ⌒
        +1     +1
        ⌒    ⌒
      +1   +2   +03
      ⌒   ⌒   ⌒
    +1  +02  +04  +07
    ⌒   ⌒   ⌒   ⌒
   5  06   08   12   19
```

Sequence 5.8..3

```
              +0
            ⌒
        +1     +1
        ⌒    ⌒
      +3   +4   +05
      ⌒   ⌒   ⌒
    +1  +04  +08  +13
    ⌒   ⌒   ⌒   ⌒
   9  10   14   22   35
```

Sequence 5.8..4

4.9.1 Nth TERM OF PENTANUMERIC PROGRESSION

This time try to do the whole process by yourself. Write the skeleton structure 1st. Then think about the terms, how terms are appearing. What will you do to get T_2, T_3, T_4 … and so on. Write at least an expression for about 10 terms to have a clear vision about the pattern. Can you find any pattern after writing? If yes then well enough or else go with me.

```
              +0      +0    +0        +0
           +A₂   +A₂   +A₂   +A₂   +A₂
         A₃    Z₂    Z₃    Z₄    Z₅    Z₆
       A₄    Y₂    Y₃    Y₄    Y₅    Y₆    Y₇
     A₅    T₂    T₃    T₄    T₅    T₆    T₇    T₈
```

Skelton structure of pentanumeric progression

As usual we will use

A_5 for the first term of Pentanumeric progression.

A_4 for the first term of Fourth slope

A_3 for the first term of the third slope

D or A_2 for the common difference.

And

$Z_2 = A_3 + A_2$

$Z_3 = A_3 + 2A_2$

$Z_4 = A_3 + 3A_2$

$Z_5 = A_3 + 3A_2$

$Z_6 = A_3 + 4A_2$

$Y_2 = A_4 + A_3$

$Y_3 = A_4 + 2A_3 + d$

$Y_4 = A_4 + 3A_3 + 3d$

$Y_5 = A_4 + 4A_3 + 6d$

$Y_6 = A_4 + 5A_3 + 10d$

Now,

The Pentanumeric progression can be expressed as

For,

$T_1 = A_5$

$T_2 = A_5 + A_4$

$T_3 = T_2 + Y_2 = A_5 + 2A_4 + A_3$

$T_4 = T_3 + Y_3 = A_5 + 3A_4 + 3A_3 + D$

$T_5 = T_4 + Y_4 = A_5 + 4A_4 + 6A_3 + 4D$

$T_6 = T_5 + Y_5 = A_5 + 5A_4 + 10A_3 + 10D$

Here coefficient of A_5 is common in all, coefficient of A_4 can be expresssed as (n-1), coefficient of A_3 forms the progression 0,0,1,3,6,10,15,... whose nth term is $\frac{(n-2)(n-1)}{2}$

And coefficient of D forms the progression 0,0,0,1,4,10,20,... (Tetrahedral numbers beginning after two zeroes extra) that's why we'll subtract 2 from n.

S(n)=

$S(n-1) = \frac{(n-3)(n-2)(n-1)}{3!}$

And whole can be Rewrite as

$A_5 + (n-1)A_4 + \frac{(n-2)(n-1)A_3}{2} + \frac{(n-3)(n-2)(n-1)D}{3!}$

Or

$T_n = A_5 + (n-1)A_4 + \frac{(n-2)(n-1)}{2}\left[A_3 + \frac{(n-3)D}{3}\right]$

4.9.2 n^{th} TERM FROM END

We can move in any direction we want, throughout a progression. We can go from first to ahead, middle to ahead or back and from last to first. Even If we reverse the order i.e. Designate the last term as first and first as last, the

NUMERIC PROGRESSION

progression will still maintain its pentanumeric nature. We will use this concept here to find the nth term from the end of the pentanumeric progression. Consider the Skelton structure of pentanumeric progression.

$$
\begin{array}{cccccccc}
 & & +0 & +0 & +0 & +0 & & \\
 & +A_2 & +A_2 & +A_2 & +A_2 & +A_2 & & \\
 & A_3 & Z_2 & Z_3 & Z_4 & Z_5 & Z_6 & \\
 A_4 & Y_2 & Y_3 & Y_4 & Y_5 & Y_6 & Y_7 & \\
 A_5 & T_2 & T_3 & T_4 & T_5 & T_6 & T_7 & T_8 \\
\end{array}
$$

Rewrite it as $T_8, T_7, T_6, T_5, T_4, T_3\ldots$ and now explain it from the bottom like this

$$
\begin{array}{cccccccc}
 & -Y_7 & -Y_6 & -Y_5 & -Y_4 & -Y_3 & -Y_2 & -A_4 \\
T_8 & T_7 & T_6 & T_5 & T_4 & T_3 & T_2 & A_5 \\
\end{array}
$$

A Complete explanation is below

$$
\begin{array}{cccccccc}
 & & +0 & +0 & +0 & +0 & & \\
 & -A_2 & -A_2 & -A_2 & -A_2 & -A_2 & & \\
 & +Z_6 & +Z_5 & +Z_4 & +Z_3 & +Z_2 & +A_3 & \\
 & -Y_7 & -Y_6 & -Y_5 & -Y_4 & -Y_3 & -Y_2 & -A_4 \\
T_8 & T_7 & T_6 & T_5 & T_4 & T_3 & T_2 & A_5 \\
\end{array}
$$

53

Where,

A_5 for the first term of Pentanumeric progression.

A_4 for the first term of Fourth slope

A_3 for the first term of the third slope

D or A_2 for the common difference.

$Z_2 = A_3 + A_2$

$Z_3 = A_3 + 2\,A_2$

$Z_4 = A_3 + 3\,A_2$

$Z_5 = A_3 + 3\,A_2$

$Z_6 = A_3 + 4\,A_2$

$Y_2 = A_4 + A_3$

$Y_3 = A_4 + 2\,A_3 + d$

$Y_4 = A_4 + 3\,A_3 + 3d$

$Y_5 = A_4 + 4\,A_3 + 6d$

$Y_6 = A_4 + 5\,A_3 + 10d$

General formula for nth term of pentanumeric progression is

$$T_n = A_5 + (n-1)A_4 + \frac{(n-2)(n-1)A_3}{2} + \frac{(n-3)(n-2)(n-1)D}{3!}$$

NUMERIC PROGRESSION

The second and fourth slope is negative. The value of A_4 (first term of slope 4) and A_3 (First term of slope 3) has changed because we had reversed its order. But Remember slope 4 is still Tetranumeric progression and slope 3 is still arithmetic progression. Lets place -# in place of the first term of slope 4 and α in place of the first term of slope 3. Since all terms of slope 4 and 2 are negative we'll place – sign before it.

$$T_n = A_5 + (n-1)(-\#) + \frac{(n-2)(n-1)\alpha}{2} - \frac{(n-3)(n-2)(n-1)(-d)}{3!}$$

As we know from earlier that as we go up, the number of terms decreases by 1. So if the pentanumeric progression has 8 terms, slope 4 will contain 7 terms i.e.(N-1) terms and slope 3 will 6 i.e. (N-2) terms.

Tn for tetranumeric progression is

$$T_n = A_4 + (N-1)\left[A_3 + \frac{(N-2)D}{2}\right]$$

Where N is the given number of terms of tetranumeric progression. Here in the case, it is N-1 with respect to given terms of pentanumeric progression.

T_n(new) for tetranumeric progression that is compatible with pentanumeric is

$\# = A_4 + (N-2)[A_3 + \dfrac{(N-3)D}{2}]$

Now A_3, which is the first term of an arithmetic progression.

The general formula of AP is

$A_3 + (N-1)d$

Since the value of N is two less than the original number of terms in pentanumeric progression. We will subtract two from it.

$A = A_3 + (N-3)d$

Now replacing the new formulaes and sign in original formula of nth term of pentanumeric progression, we will get

T_n from end =

$T_n = A_5 - (n-1)[A_4 + (N-2)\{A_3 + \dfrac{(N-3)D}{2}\}] + \dfrac{(n-2)(n-1)}{2}[A_3 + (N-3)d] - \dfrac{(n-3)(n-2)(n-1)D}{3!}$

Or

$L - (n-1)[A_4 + (N-2)\{A_3 + \dfrac{(N-3)d}{2}\}] + \dfrac{(n-2)(n-1)}{2}[A_3 + (N-3)d - \dfrac{(N-3)D}{3}]$

NUMERIC PROGRESSION

Let the progresssion 2,10,20,33,50,…. And explain it.

$$
\begin{array}{cccccc}
 & & +1 & & +1 & \\
 & +2 & & +3 & & +4 \\
 & +8 & +10 & & +13 & +17 \\
2 & & 10 & 20 & 33 & 50
\end{array}
$$

It will be rewrited and explained as

$$
\begin{array}{ccccccc}
 & & -1 & & -1 & & -1 \\
 & +4 & & +3 & & +2 & & +1 \\
 & -17 & & -13 & & -10 & & -8 & -7 \\
50 & & 33 & & 20 & & 10 & & 2 & -5
\end{array}
$$

Fig- Example of conversion

4.9.3 SUM OF Nth TERM

Once again consider the Skelton structure of pentanumeric progression.

$$
\begin{array}{ccccc}
+0 & +0 & +0 & +0 \\
+A_2 & +A_2 & +A_2 & +A_2 & +A_2 \\
A_3 & Z_2 & Z_3 & Z_4 & Z_5 & Z_6 \\
A_4 & Y_2 & Y_3 & Y_4 & Y_5 & Y_6 & Y_7 \\
A_5 & T_2 & T_3 & T_4 & T_5 & T_6 & T_7 & T_8
\end{array}
$$

Where,

A_5 for the first term of Pentanumeric progression.

A_4 for the first term of Fourth slope

A_3 for the first term of the third slope

D or A_2 for the common difference.

$Z_2 = A_3 + A_2$

$Z_3 = A_3 + 2A_2$

$Z_4 = A_3 + 3A_2$

$Z_5 = A_3 + 3A_2$

$Z_6 = A_3 + 4A_2$

$Y_2 = A_4 + A_3$

$Y_3 = A_4 + 2A_3 + d$

$Y_4 = A_4 + 3A_3 + 3d$

$Y_5 = A_4 + 3A_3 + 6d$

$Y_6 = A_4 + 4A_3 + 10d$

so we can express sum as

For,

$s_1 = A_5$

$s_2 = A_5 + T_2 = 2A_5 + A_4$

$s_3 = A_5 + T_2 + T_3 = 3A_5 + 3A_4 + A_3$

$s_4 = A_5 + T_2 + T_3 + T_4 = 4A_5 + 6A_4 + 4A_3 + D$

Same wise

$s_5 = 5A_5 + 10A_4 + 10A_3 + 5D$

$s_6 = 6A_5 + 15A_4 + 20A_3 + 15D$

$s_7 = 7A_5 + 21A_4 + 35A_3 + 35D$

$s_8 = 8A_5 + 28A_4 + 56A_3 + 70D$

And so on...

Here, the Coefficient of A_5 is equal to n, the Coefficient of A_4 appears as 0,1,3,6,10,15,... whose nth term is $\dfrac{(n-1)n}{2}$

Coefficient of A_3 appears as 0,0,1,4,10,20,35,... whose nth term is $\dfrac{(n-2)(n-1)n}{6}$ And Coefficient of d appears as 0,0,0,1,5,15,35,70,... whose nth term is $\dfrac{(n-3)(n-2)(n-1)n}{24}$

By assembling these we shall get the general formula for Sum of the nth term of Pentanumeric progression

$$S_n = nA_5 + \dfrac{(n-1)nA_4}{2} + \dfrac{(n-2)(n-1)nA_3}{3!} + \dfrac{(n-3)(n-2)(n-1)nD}{4!}$$

Or

$$S_n = \dfrac{24nA_5 + (n-1)12nA_4 + (n-2)(n-1)4nA_3 + (n-3)(n-2)(n-1)nD}{24}$$

Or

$$S_n = \dfrac{1}{24}[24nA_5 + (n-1)12nA_4 + (n-2)(n-1)4nA_3 + (n-3)(n-2)(n-1)nD]$$

4.10 HEXANUMERIC PROGRESSION

We will do exactly the same as so far we are doing with the previous ones. We are at the edge to find out the general formula for Numeric progression. Perform the same steps with hexanumeric.

Some Explained Hexanumeric progression

```
                  + 1     +1
             + 2      +3      + 4
         + 8    + 10   + 13   +17
       2    10     20     33     50
    1    03    13    33    66    116
```

```
                  + 1     +1
             + 1      +2      + 3
         + 4    + 05   + 07   +10
       2    06     11     19     29
    2    04    10    21    40    69
```

we'll get the following:-

The Nth term of hexanumeric progression

$$T_n = A_6 + (n-1)A_5 +$$

$$\frac{(n-2)(n-1)A_4}{2} + \frac{(n-3)(n-2)(n-1)A_3}{6}$$

$$+ \frac{(n-4)(n-3)(n-2)(n-1)d}{24}$$

$$T_n \text{ from end}$$

$$= L - (n-1)[A_5 + (N-2)A_4 +$$

$$\frac{(N-3)(n-2)}{2}\{A_3 + \frac{(N-4)D}{3}\}] + \frac{(n-2)(n-1)}{2}[A_4 + (N$$

$$- 3)\{A_3 + \frac{(N-4)D}{2}\}] - \frac{(n-3)(n-2)(n-1)}{6}\{A_3$$

$$+ (N-4)D\} + \frac{(n-4)(n-3)(n-2)(n-1)d}{24}$$

Sum of nth term of hexanumeric progression

$$S_n = nA_6 + \frac{(n-1)nA_5}{2} + \frac{(n-2)(n-1)nA_4}{3!} +$$

$$\frac{(n-3)(n-2)(n-1)nA_3}{4!} +$$

$$\frac{(n-4)(n-3)(n-2)(n-1)nd}{5!}$$

5. GENERAL FORMULA

5.1 General formula for the nth term of numeric progression

So far, we are finding the formulae one by one. Here we will have a combined study of all those formulae to find a common pattern between them. Let's list out all the formula for the nth term of individuals progressions.

Common arithmetic progression = A_2

Arithmetic progression = $A_3 + (n-1)d$

Tetranumeric progression = $A_4 + (n-1)A_2 + \dfrac{(n-2)(n-1)D}{2}$.

Pentanumeric progression = $A_5 + (n-1)A_4 + \dfrac{(n-2)(n-1)A_3}{2} + \dfrac{(n-3)(n-2)(n-1)D}{3!}$

Hexanumeric progression = $A_6 + (n-1)A_5 + \dfrac{(n-2)(n-1)A_4}{2} + \dfrac{(n-3)(n-2)(n-1)A_3}{6} + \dfrac{(n-4)(n-3)(n-2)(n-1)d}{24}$

Here we observe, Each formula starts with the first term of respective principle progression. Then it goes through the first term of each preceding slope up to common difference with a variable coefficient.

Let us assume a progression whose first term is A_n

Then $T_n = A_n + (n-1)A_{n-1} + \dfrac{(n-2)(n-1)A_{n-2}}{2} + \dfrac{(n-3)(n-2)(n-1)A_{n-3}}{3!} + \dfrac{(n-4)(n-3)(n-2)(n-1)A_{n-4}}{4!} + \dfrac{(n-5)(n-4)(n-3)(n-2)(n-1)A_{n-5}}{5!} + \ldots\ldots$

The process will continue until A_{n-z} reaches common difference. Or $A_{n-z} = A_2$ Where z is the whole number.

What we also observe is:-

Equation of Arithmetic progression is linear in nature,

Equation of Tetranumeric progression is quadratic in nature,

Equation of Pentanumeric progression is cubic in nature, and so on…

Also, vice -versa is true i.e.

Linear expression gives births to Arithmetic progression,

Quadratic expression gives tetranumeric progression,

Cubic expression gives pentanumeric progression and so on…

5.2 General formula for the nth term from the end of a numeric progression

Arithmetic progression $= L-(n-1)d$

Tetranumeric progression $= L-(n-1)\{A_2+(N-2)d\} + \frac{(n-2)(n-1)D}{2}$

Pentanumeric progression $= L-(n-1)[A_4+(N-2)\{A_3+\frac{(N-3)d}{2}\}] + \frac{(n-2)(n-1)}{2}[A_3+(N-3)d - \frac{(n-1)(n-2)(n-3)D}{3!}$

Hexanumeric progression =

$= L-(n-1)[A_5+(N-2)A_4 + \frac{(N-3)(n-2)}{2}\{A_3 + \frac{(N-4)D}{3}\}] + \frac{(n-2)(n-1)}{2}[A_4 + (N-3)\{A_3 + \frac{(N-4)D}{2}\}] - \frac{(n-3)(n-2)(n-1)}{6}\{A_3 + (N-4)D\} + \frac{(n-4)(n-3)(n-2)(n-1)d}{24}$

General formula of numeric progression is

$T_n = A_n + (n-1)A_{n-1} + \frac{(n-2)(n-1)A_{n-2}}{2} + \frac{(n-3)(n-2)(n-1)A_{n-3}}{3!} +$

$$\frac{(n-4)(n-3)(n-2)(n-1)A_{n-4}}{4!}$$

$$+ \frac{(n-5)(n-4)(n-3)(n-2)(n-1)A_{n-5}}{5!} + \ldots$$

....

Here we can observe that

The sign changes alternately i.e. L is always positive and then it goes negative, then positive and so on.... The variation continues until we reach d.

We have to replace the value of the first term of each slope with the formula of each respective slope and then we have to subtract one from n of general formula of the first term of principle progression and then 2 from N of the general formula for one slope lesser after it subtracts 3 then 4 until it reaches d.

Let us assume a progression whose end term is L and progression contain Z slope

$L - (n-1)(\text{general formula of Z slope with oneless in } N) + \frac{(n-2)(n-1)}{2} [\text{general formula of } Z-1 \text{ slope with two less in } N]$

$- \frac{(n-3)(n-2)(n-1)}{3!} \{\text{general formula of } Z-2 \text{ slope with three less in } N\}$

$+ \frac{(n-4)(n-3)(n-2)(n-1)(\text{general formula of } Z-3 \text{ slope with four less in } N)}{4!}$

-;..................................

Until Z-K slope coincides with D. where k starts from zero and progresses by one with the Z.

5.3 General formula for the sum of the nth term of numeric progression

Arithmetic progression $= nA_3 + \dfrac{n(n-1)D}{2}$

Tetranumeric progression $= nA_4 + \dfrac{n(n-1)A_2}{2!} + \dfrac{(n-2)(n-1)nD}{3!}$

Pentanumeric progression $= nA_5 + \dfrac{(n-1)nA_4}{2} + \dfrac{(n-2)(n-1)nA_3}{3!} + \dfrac{(n-3)(n-2)(n-1)nD}{4!}$

Hexanumeric progression $= nA_6 + \dfrac{(n-1)nA_5}{2} + \dfrac{(n-2)(n-1)nA_4}{3!} + \dfrac{(n-3)(n-2)(n-1)nA_3}{4!} + \dfrac{(n-4)(n-3)(n-2)(n-1)nd}{5!}$

Here we observe that

n is multiplied in the first term of respective progression. And then it goes through each first term of different preceding slope up to common difference with a variable multiplied in each first term.

Let us assume a progression whose first term of principle slope is A_n

Then $s_n = nA_n + \dfrac{(n-1)nA_{n-1}}{2} + \dfrac{(n-2)(n-1)nA_{n-2}}{3!} + \dfrac{(n-3)(n-2)(n-1)nA_{n-3}}{4!} + \dfrac{(n-4)(n-3)(n-2)(n-1)nA_{(n-4)}}{5!} + \ldots\ldots\ldots$

Up to A_{n-z} coincides with D. Where Z is some integer.

Author's words

This is not the end of the numeric progression rather than this, it is starting of your journey with numeric progression and other types of sequences. We start with some common sequences and went through several sequences. After reading this you must be searching for sequences near you. My journey with sequences also started from observation. Whether it is a ranking system in my class or increment in my pocket money. This is all very much fascinating. The first time, I begin observing numeric progression in my high school. it takes me a year to observe this topic and writing the stuff. And I am very honored to share all of this with you. If you find one, don't forget to share it with me.
mailto:purushottamkumarsuman1002@gmail.com
Remember it is surrounding you. Also, don't forget to share your reviews or comments. I am much more waiting for it. Thank you for reading.

Best of luck for the journey!

www.ingramcontent.com/pod-product-compliance
Lightning Source LLC
Chambersburg PA
CBHW021501210526
45463CB00002B/830